# Social Media is Fake

*How Bots, Algorithms, and AI Are Replacing Real Life*

by

## BRIAN TURNER

For permissions, inquiries, or bulk orders, please contact:
 hi@heybbt.com
 www.heybbt.com

First Edition
ISBN:  979-8-9931382-7-5

Printed in the United States of America

# Table of Contents

# Intro: The Illusion of Connection

*Social media broke us. We don't scroll to connect anymore. We scroll to forget.*

*Scroll. Like. Share. Repeat.*

That's the ritual: wake up, grab the phone, and step into the feed. It feels alive, buzzing, endless, full of voices. But most of what you are looking at is not real.

Nearly half of all internet traffic today comes from bots. Not people. Not your friends. Not fans. Just code programmed to mimic human behavior and feed you what keeps you scrolling.

Platforms know it. Twitter (now X) admitted that 9 to 15 percent of its accounts are bots, and researchers say in political debates, that number can climb toward 60 percent. Facebook admitted to more than 80 million fake accounts. Instagram is a jungle of engagement pods, fake likes, and comment farms.

Now add AI. Fake faces. Fake influencers. Fake arguments. Entire accounts that look and sound human but are nothing but code. There are AI influencers with millions of followers who do not even exist. And here's the wild part: people trust them more than they trust real creators.

## The Moment We Went Numb

Do you remember the first time you saw something shocking online? Maybe it was a fight, a death, or a tragedy caught on camera. You felt something. You paused.

Now? We scroll past murders the same way we scroll past memes, barely noticing either.

We laugh at things that should make us cry. We cheer for things that should make us pause. We treat human life as if it were just another piece of content, sandwiched between food pictures and dance videos.

We're not bad people. We're just numb. And the feed trained us to be this way.

## A Broken Mirror

Social media is supposed to reflect society, but what happens when the mirror is cracked?

The feed does not show us the truth. It shows us what will keep us there the longest. Outrage, not nuance. Anger, not honesty. False news spreads faster than facts. Bots boost the worst voices. Algorithms feed us poison disguised as connection, and bots boost the worst voices. We're no longer looking at life. We're looking at a simulation.

---

## The Kids Are Watching

I think about my kids. They're growing up in a world where fake faces get more love than real ones. Where a bot can win an argument against a human. Where attention is more important than empathy.

What kind of world have we built?

And one day they will have to ask the question we are too scared to face: *what part of this is real?*

---

## Why This Book Exists

This book is not about hating technology. It is about seeing it clearly. The scariest part is not that social media is fake. It is that we got comfortable with it.

I am convinced something has to change. Either the platforms change, or we do. Because if we keep going like this, the line between real and fake disappears, and once it is gone, there is no bringing it back.

---

## Closing One-Liner

*We did not lose ourselves to social media. We lost ourselves to the version of ourselves we thought people wanted to see.*

# Chapter 1 – The Dead Internet Theory

*Half the internet is not human. We built
a digital city, then filled it with ghosts.*

The first time I heard about the "Dead Internet
Theory," I laughed. It sounded crazy. The claim was
that most of what you see online isn't people at all.
The posts, videos, comments, and even accounts are
all automated. Bots running the show.

It started showing up on message boards back in
2018. Folks saying the internet had flipped. Still loud
on the surface. Still busy. Still buzzing. But
underneath? Empty.

What looked like millions of people talking were
really scripts arguing with scripts. What looked like
trends rising from the crowd were machines
planting seeds and juicing them until they felt real.

The more you sit with it, the harder it is to ignore.
Why does the internet feel like the same memes on
repeat? Why do half the replies under a post make
no sense, like nobody's actually there? Why does
going viral feel bigger than it should, like the
cheering crowd is pumped in through the
speakers?

That's the Dead Internet Theory in plain terms. The
idea that what you scroll every day isn't a town

square. It's a stage play. The people are props. The noise is scripted. And you're the only one in the audience who doesn't realize it's fake.

It's *The Truman Show*, but instead of one man trapped inside a fake world, it's all of us. Whole societies are stuck inside a feed. Every like, every comment, every "trending topic" is carefully placed to make it feel alive.

And just like Truman, most people never question it. They laugh, they cry, they react like it's real. But the set was built. The script was written. The crowd was planted.

The only difference is that in *The Truman Show*, once you saw the walls, you could walk out. Online? The walls are invisible. And the show never ends.

The internet isn't crowded anymore. It just feels that way because the ghosts won't shut up.

And here's the kicker: it might not even be a conspiracy.

In 2016, researchers said more than half of all web traffic was bots. By 2023, it was still close to half. Half the clicks. Half the comments. Half the activity you see online might not be human at all.

Think about that. You could be arguing with a ghost. You might be liking a post that's been juiced up by code. You could be sharing something that never came from a person in the first place.

## The Dead Internet in Plain Sight

And this is not just obscure corners of the web. It's baked into the biggest platforms.

Twitter was once estimated to have tens of millions of bot accounts active in a single year. That meant a large chunk of "users" weren't real people at all, and worse, bots post far more than humans, so they dominated timelines.

In 2019, Facebook revealed it had disabled more than three billion fake accounts in just six months. That's more fake profiles than the population of China.

TikTok has been caught with massive "engagement farms" in Southeast Asia, where rows of phones run 24/7 to inflate likes and views. For the right price, anyone can look like a star.

And in political debates, bots often made up less than one percent of users but generated nearly a third of the content. Imagine walking into a stadium with a thousand people. Only ten are bots, but they've got megaphones that never turn off. That's the internet.

We're not logging into a digital town square. We're walking into a masquerade ball where half the masks are plastic.

## The Viral Feeling That Isn't Real

Here's what makes it relatable: if you've ever had a post blow-up online, you know the feeling. Your phone explodes with likes, comments, and new followers. At first, it feels amazing. But then you start scrolling through the replies. Half of them make no sense. Some are copied and pasted. Some look like real people, but their profiles are empty or stolen.

That's when it hits you. This "viral moment" that felt like everyone was watching? A lot of it was fake. Much of it involved machines pretending to be people.

And yet, the platforms count it all the same. To them, a bot view is just as valuable as a human view. Engagement is engagement, even if it's hollow.

## Why It Matters

The danger isn't just the bots themselves. It's what they do to our sense of reality. Bots aren't neutral. They're programmed to push agendas, flood conversations, and distort what feels popular.

We don't debate ideas online anymore. We debate machines pretending to be people.

That's why false information spreads faster than the truth. That's why outrage dominates your feed. That's why certain voices look louder than they really are. Bots tilt the stage, and the algorithm rewards them.

Makes you wonder, if half the internet is fake, how much of what we believe is shaped by ghosts?

---

## The Human Cost

This isn't just about numbers. It changes how we live.

Take #BlackLivesMatter. The movement was real, built on pain and urgency. But bots hijacked it, flooding feeds with fake accounts posting extreme content on both sides. Some pretended to be activists. Others pretended to be police supporters. Both were run by the same networks. The goal wasn't justice. The goal was division.

That's what bots do best. They take something real, something human, and they warp it. They make us fight each other while they amplify the chaos.

And while we fight shadows, the platforms profit. More engagement. More time spent. More ad dollars.

The cost is that we no longer know what voices are real. The cost is that we no longer know which voices are real. And when you can't tell who's real, you stop trusting anyone.

---

## Closing One-Liner

> *The internet is not dead. It's undead running on ghosts that look human enough to fool us.*

# Chapter 2 – The Rise of Social Media Bots

*You didn't lose the argument online. You lost it to a bot programmed to make you mad.*

Bots didn't start out running the internet. At first, they were clumsy. A script that liked a post. A fake account with no photo. A spam comment on YouTube saying "Nice video!" Background noise. Easy to spot. Easy to ignore.

But bots evolved. They got smarter. Harder to catch.

Now they build full personalities. They steal profile pictures. They post memes. They reply with jokes. They argue with each other, so a thread looks alive. Some even send you DMs like they're human.

Once bots started looking and sounding real, they stopped being noise. They started shaping the feed.

---

## The Timeline of Fakes

### Stage One: Boosters
Early bots inflated numbers. Auto-follows on Twitter. Auto-likes on Instagram. Fake views on YouTube. Their only job was to make someone look bigger than they were.

## Stage Two: Agitators

Then came bots built to stir fights. They dropped comments to start drama. They hijacked hashtags: politics, sports, celebrity gossip. Bots showed up not to win, but to drown everyone else out.

## Stage Three: Salesmen

E-commerce jumped in. Fake reviews on Amazon. Copy-paste comments pushing products. Whole sites that exist just to pump five-star reviews. Enough to make you trust a product nobody ever really touched.

One of the clearest examples came when selfie sticks flooded Amazon. Out of nowhere, dozens of no-name brands showed up with thousands of glowing reviews. People thought everyone was buying them. Most of those reviews were fake, paid posts. Review rings. The sticks were cheap junk. The hype was the product.

That's how it works. A seemingly random gadget, supplement, or workout tool suddenly becomes the next big thing. Not because it's good, but because bots made it look that way.

## Stage Four: Impersonators

Now bots blur the line between fake and real. AI writes better captions. Stolen profile photos create fake faces. Some accounts you scroll past every day — the ones you think are people — are nothing but scripts pretending to live lives.

## The Sting

This is the part we don't admit. We've already been fooled.

That faceless account that argued with you for hours. That Amazon review that convinced you to click "Buy." That influencer with a million followers was later exposed because half of their followers were fake. That rapper who got caught because their streaming numbers were impossible in real life.

We don't argue with people online. We argue with scripts trained to waste our time.

## The Human Cost

Bots don't just pad numbers. They shape culture.

In politics, they hijack debates. They boost extremes until the middle disappears. They do not just spread lies. They spread outrage, because outrage spreads faster than truth.

In music and entertainment, they decide who looks popular. Rappers have been caught juicing streams with bots. Influencers buy fake followers to lock in brand deals. Whole industries pay people to run bot farms so you believe the crowd is real.

And in everyday life, fake has become normal. Influencers buy bots to look famous. Musicians boost streams to climb charts. Brands pay for fake reviews because real ones aren't enough.

We have built an economy on fake engagement. Here's the gut punch: most of us spend hours making content just to feed that same fake system.

We think we're building something. Most of the time, we're just giving bots more fuel.

---

**Closing One-Liner**

> *The bots aren't hiding anymore. They run the feed. And most of us can't tell the difference.*

# Chapter 3 – Manufactured Virality

*Nothing "just went viral." Someone made it happen.*

We like to believe the internet is magic. That if something is good enough, funny enough, or shocking enough, it will go viral. That the crowd decides.

That's a lie.

Virality is built. It's paid for, boosted, and planted. If you see something everywhere at once, it's because someone put it there.

A new song drops, and suddenly it's on TikTok, Instagram Reels, and YouTube Shorts at the same time. You think the world just discovered it overnight. What really happened? Labels cut checks. Influencers received scripts that told them exactly how to use the track. The algorithm was primed to push it. The crowd did not choose it. The crowd was led to it.

Same with memes. A random meme doesn't take over the internet because people happen to share it. There are pages, networks, and bot accounts built to push it until it looks organic. By the time you laugh at it, it's already been manufactured for you.

**The Playbook**

1. **Seed it.** Pay a few accounts with reach. Drop it in group chats, forums, or pages.

2. **Boost it.** Use bots or networks to inflate early likes and comments.

3. **Flood it.** Once it looks like it's catching, push it across platforms.

4. **Claim it's viral.** Articles, blogs, and even news outlets will start calling it the internet's new obsession.

At that point, real people join in. Not because it is authentic, but because it already looks popular.

That's how we got things like the TikTok Ocean Spray skateboard video. A guy posts a clip riding a longboard, drinking juice, vibing to Fleetwood Mac. It felt random and pure. The video went everywhere. Ocean Spray sent him a truck. Fleetwood Mac charted again. But behind the curtain, TikTok's algorithm had already decided to push it, and brands jumped in once they smelled money. It looked organic. It was orchestrated.

Same with the Harlem Shake. Out of nowhere, everyone on Earth seemed to be doing it. Schools.

Offices. Celebrities. It looked like a grassroots explosion. In reality, networks and media pushed it until it felt unstoppable.

And remember the Popeyes chicken sandwich in 2019. One tweet turned into nationwide hype. Lines around the block. Sold out everywhere. It felt like culture. It was marketing dressed up as a movement.

And I know this firsthand. I went viral once.

It didn't feel like success. It felt fake. It felt heavy.

One day, the room was quiet. The next day, my phone wouldn't stop buzzing. Notifications stacked like they meant something. But half the comments felt hollow.

That's when it hit me. This game is rigged.

You can post something powerful, and nobody sees it. Then you boost the same thing, and suddenly everyone loves it. Even your own followers don't see your work until the platform decides they should.

---

**The Sting**

We think the internet is a democracy. It isn't. It's a stage show with the spotlight rented out.

We're not discovering trends. We're being handed them.

## The Human Cost

Manufactured virality is why nothing feels real anymore. Songs don't blow up because they're good. Products don't trend because they work. People don't go viral because they're interesting.

It's marketing dressed up as magic.

And it does something worse to us. It makes us feel like if we're not going viral, we're failing. Like we're invisible. Like we don't matter.

So we copy the formula. We chase the algorithm. We waste hours trying to package our lives into the next "viral moment."

But even when it works, it's still fake. You can "go viral" and never know how much of it was bots, boosts, or money behind the curtain.

## Closing One-Liner

> Virality is not luck. It's leverage disguised as magic.

# Chapter 4 – The Engagement Machine

*The real influencer isn't in your feed. It is the algorithm you will never meet.*

We think influencers run the internet. They don't. The algorithm does.

The algorithm decides who gets seen. Who gets buried. What trends. What dies.

And it has one goal: keep you scrolling.

It doesn't care about truth. It doesn't care about quality. It doesn't care about you. It cares about engagement.

---

## How It Feeds

The machine feeds on signals. Likes. Comments. Shares. Watch time. Saves. Every tap and pause is fuel.

The more you react, the more it learns. The more it learns, the more it feeds.

It doesn't care if you're laughing, arguing, or crying. It just wants you to stay. Outrage works. Controversy works. Conflict works.

That's why your feed feels heavier than real life because the algorithm rewards what gets a reaction, not what makes you better.

Instagram knew this. Leaked reports showed how the platform boosted posts with heated comments because fights keep people scrolling. The more people argued, the more the post was shown. Anger became a growth strategy.

TikTok took it further. Internal reports revealed the company "whitelisted" certain creators, meaning their videos were pushed to millions of people on purpose. It looked like they blew up overnight. In reality, the platform decided who got famous.

And Facebook? For years, its algorithm quietly boosted political misinformation because outrage kept people glued to their screens. The company initially denied it, but later admitted it after whistleblowers came forward. The truth didn't matter. Only clicks did.

---

**The Sting**

The algorithm isn't showing you the world. It is showing you what will keep you stuck.

---

**The Human Cost**

We used to think we were chasing connection. Really, we were training a machine to control what we see.

It's why lies spread faster than facts. Why fights trend faster than solutions. Why you know more about celebrity drama than your own neighbors.

The algorithm isn't neutral. It tilts the table. It makes anger look normal. It makes division look popular.

And the worst part? We play along. We shape our content to please the machine. We post for engagement instead of meaning. We bend our voices until the algorithm approves.

I know this firsthand. For a minute, I was having some success on YouTube. A few posts came with decent-sized checks—all organic. No ads. No boosts.

I learned something quickly. If you chase the algorithm, the traffic will come.

But I stopped.

First, I got flagged for talking about Alex Jones and Ye. Second, I was losing myself.

The money was there. The views were there. But me? I was disappearing. And I knew I couldn't keep that up.

Creators burn out trying to feed it. Regular people feel invisible when they don't show their posts.

Whole movements live or die depending on what the algorithm decides to boost.

---

## Closing One-Liner

*The feed isn't built to show you life. It is built to keep you locked inside the machine.*

# Chapter 5 – Fake News, Real Consequences

*Fake news doesn't just fool strangers. It ruins families. It breaks friendships. It makes us question people we thought we knew.*

We used to think fake news was background noise. Spam posts. Conspiracy blogs. Junk articles you'd ignore and move on from.

Now it shows up at the dinner table. It shows up in text threads that end with someone leaving the chat. It shows up in arguments with people you love.

The cost isn't just confusion. It's trust.

---

## When It Gets Real

Fake news isn't harmless. It shapes what we believe, what we fear, what we fight over.

Elections tilt when people vote based on stories that never happened. Whole movements are warped by fake headlines.

The pandemic made it worse. Fake cures. Fake statistics. Fake science. Millions believed them.

Millions acted on them. Some died because of them.

And it's not always national. Sometimes it's personal. Families split over conspiracy videos. Friend groups collapse over Facebook posts. Marriages crack under the weight of what someone read online.

---

**Cultural Hits**

Remember Pizzagate?

In 2016, a story spread online claiming a Washington, D.C. pizza shop was running a secret child trafficking ring tied to politicians. Reddit threads blew up. Twitter accounts pushed it. Conspiracy blogs wrote post after post.

It was all fake.

But it didn't stay online. A man from North Carolina drove six hours with an AR-15 to investigate the incident. He walked into the restaurant, pointed his gun, and demanded answers. He thought he was there to save kids from a basement that never existed.

Nobody died that day. But it showed how far fake news can go. A rumor on the internet turned into a man with a gun in a real restaurant. That's the world we live in now.

Or take the Tide Pod Challenge. It started as a joke, but headlines turned it into a moral panic. Parents thought kids everywhere were eating laundry pods. Companies scrambled. Schools lectured. A fake crisis was treated as if it were life or death.

Fake spreads fast. And when people believe it, the consequences land in the real world.

---

## The Desensitization Trap

The phrase fake news became a punchline. At first it meant dangerous misinformation. Then it turned into a label people slapped on anything they didn't like. It became a meme. A way to end an argument. Background noise.

And that's the trap. The louder the world yelled fake news, the less people took it seriously. The more it sounded like a joke, the easier it was to ignore.

But fake news itself never stopped. The lies still spread. The rumors still shaped elections. The conspiracies still broke families apart. We got desensitized to the words, but the consequences stayed real.

---

## The Sting

Fake news isn't entertainment. It's a weapon.

## The Human Cost

Fake news divides. It isolates. It breaks down trust.

It turns neighbors into enemies. It makes families stop talking. It convinces people to believe wild stories while ignoring real problems.

And the platforms know this. They have the data. They studied it. They saw that the crazier the headline, the longer we stayed. Outrage paid better than truth.

They could have slowed it down. They could have built guardrails. Instead, they poured gasoline on it. Because it worked.

## Closing One-Liner

> *Fake news isn't a glitch in the system. It's the system doing exactly what it was built to do.*

# Chapter 6 – The Age of Influence

*We live in a time where being an influencer is more valuable than being real.*

Influence used to mean power. Respect. A voice people listened to.

Now influence means followers. Views. Likes. Engagement.

The word got hollow. And the role became a job.

---

## The Rise of the Influencer

Social media created a new class of people. Not athletes. Not artists. Not leaders. Just people who figured out how to package their lives for the algorithm.

Some turned it into careers. They got brand deals. They launched products. They built empires.

Others built fake versions of themselves. Fake followers. Fake engagement. Fake flexes.

And the lines blurred. Who is real? Who is manufactured? Does it even matter if the numbers still add up?

---

## The Legacy Question

We don't ask this enough: what's the legacy of a social media influencer?

Thousands of posts. Millions of likes. Endless reels, stories, and shorts. But when the platforms change, when the algorithms shift, what's left?

If your profile disappeared tomorrow, what would be left?

Most influencers fade. Their fame doesn't transfer. Their audience doesn't follow. Their influence lives and dies inside a platform they don't control.

And when they pass away, the truth hits harder. Their profiles live on like ghosts. Their legacy is a highlight reel on borrowed time.

---

## Cultural Hits

An influencer was once exposed for buying almost all their followers. Brands cut ties. Their career ended overnight.

Another influencer died young, and their page became a shrine. Comments flooded in, but months later the algorithm moved on. Their life became content, recycled for engagement one last time.

Even the biggest names get caught in it. Rappers exposed for juicing streams. Celebrities buying bots to keep numbers high. Whole industries propped up by fake signals.

And then there was the influencer meltdown. A creator with millions of followers launched their own merch line. They expected thousands of sales but sold fewer than thirty. One viral thread said it best: "You don't have influence. You just have followers."

---

## The Sting

Influence online isn't about impact. It's about illusion.

---

## The Human Cost

We chase attention like it's currency. We measure our value in likes. We call it influence, but really it's performance.

People build whole identities around being influencers. They spend hours creating content, chasing numbers, begging for relevance. But most of it is smoke.

Creating content is work. Long hours. Endless edits. And even then, you have to play by the platform's rules just to be seen. No matter how good your content is, it doesn't matter if the algorithm decides it doesn't fit.

And here's the trap. The moment you start to see success, you're hooked. I hate to be dramatic, but it feels like slavery to the algorithm. You create for it. You change for it. You bend your life around what it wants.

The platforms use you. The brands use you. And when the numbers dip, when the algorithm moves on, the influencer gets left behind.

We're told influence is the new dream. But most of the time it's a trap. A job without stability. A life without privacy. A hustle built on fake numbers and rented attention..

---

**Closing One-Liner**

> *The Age of Influence isn't about being real. It's about who can fake it the longest before the curtain drops.*

# Chapter 7 – The Mental Health Crisis

*Social media doesn't just waste time. It eats away at your mind.*

We scroll like it's harmless. A quick check. A quick laugh. A quick distraction.

But it stacks. Every scroll is comparison. Every like is a test. Every notification is a hit.

We don't just use social media. We live inside it. And it changes us.

---

## The Comparison Trap

Comparison is baked into the feed.

Someone always has more. More likes. More followers. A better trip. A bigger house. A happier family.

Even if you know it's curated, it still hits. You still feel behind. You still feel less.

We measure our lives against highlight reels. And it hurts.

---

## The Anxiety Loop

The feed is built to keep you checking.

You post something and wait. Did they like it? Did they share it? Did it hit?

Your mood swings on numbers you can't control. One post blows up and you feel unstoppable. The next one tanks and you feel invisible.

It's not just stress. It's an addiction. Dopamine, when it works. Withdrawal when it doesn't.

---

## Cultural Hits

Studies show heavy social media use is linked to higher rates of depression and anxiety, especially in teens. One survey found that the average teenager spends over seven hours a day on screens, excluding schoolwork. That's almost a full-time job of scrolling.

And then came the bombshell. Internal research leaked from Instagram showed the company knew its platform worsened body image issues for teen girls. They had the data. They admitted it. And still, the feed kept running.

Influencers talk openly about burnout. About constant pressure. About feeling like a product instead of a person.

Even celebrities admit it. Famous athletes and musicians saying the same thing: "I had to step away. It was destroying me."

---

## The Sting

Social media isn't built to make you happy. It's built to keep you hooked.

---

## The Human Cost

We're more connected than ever, and lonelier than ever.

We spend hours staring at screens, but still feel unseen.

We call it community, but most of it is performance.

And the damage is quiet. People smiling in selfies while breaking inside. Teens comparing themselves to faces that are filtered and fake. Adults chasing validation from strangers while ignoring real relationships.

I know this firsthand. In my book *Aligned AF*, I wrote about deleting everyone. I had to. I was drowning. I'm an older man, and even I couldn't handle the feed while I was fighting through my own trials and tribulations.

If it almost broke me in my forties, what's it doing to kids at fourteen?

The mental health crisis isn't a side effect of social media. It's the product.

---

## Closing One-Liner

*The feed doesn't just steal your time. It steals your peace.*

# Chapter 8 – Cancel Culture and Digital Shame

*The internet doesn't forgive. It doesn't forget. It feeds.*

Cancel culture was supposed to mean accountability, to hold people responsible and shine light where it's needed.

But it turned into a blood sport.

Now it means mobs piling on. Doxxing. Digging up old posts. Demanding jobs lost. Demanding lives ruined.

---

## The Speed of Shame

It takes one clip. One screenshot. One out-of-context joke.

Overnight, someone goes from unknown to infamous. Their name trends. Their face spreads. Their life is dissected.

The crowd doesn't care about context. The crowd doesn't care about growth. The crowd wants a show.

---

## Cultural Hits

Comedians canceled for old stand-up bits. Athletes dropped for tweets they wrote as teenagers. Ordinary people fired because a video of them went viral.

Even celebrities get pulled under. Kevin Hart lost the Oscars job after old jokes resurfaced. Headlines everywhere. His reputation dragged for weeks. One gig gone and the world decided he was done.

And it's not just the famous. A woman once lost her job after a single clip of her went viral. One moment filmed by a stranger, reshared a thousand times, and her entire career was gone by morning.

Sometimes the outrage is justified. Sometimes it isn't. But once the machine starts, it doesn't stop.

---

## The Sting

Cancel culture isn't about justice. It's about entertainment.

---

## The Human Cost

The feed turns shame into content.

And shame spreads faster than success.

People are afraid to speak. Afraid to joke. Afraid to be themselves. Because the mob might come for them next.

We no longer separate the mistake from the person. We define people by their worst moment, looped and replayed forever.

If your worst moment was frozen online forever, what would it say about you?

I felt it too. There was a time when I posted on Facebook, then hid my own posts.

I didn't want to offend anyone. I didn't want to beg for validation.

So I lived in posting purgatory. Say something, then erase it. Share a thought, then bury it.

And the whole time, I felt like my posts were blacklisted anyway.

And here's the darkest part: platforms profit from it. They boost the outrage. They push the clips because nothing keeps people scrolling like someone else's downfall.

---

### Closing One-Liner

> *Cancel culture isn't the crowd holding you accountable. It's the crowd holding you down for clicks.*

# Chapter 9 – The Business of Fake

*Fakery isn't a glitch. It's the business plan.*

The internet sells attention. That's the product.

And nothing grabs attention like fake. Fake engagement. Fake controversy. Fake outrage.

Bots inflate the numbers. Brands chase the numbers. Platforms sell the numbers back to advertisers.

It looks like connection. It's commerce.

---

**How Platforms Profit**

Every like, every share, every second you scroll is money.

Platforms don't care if the account is real or fake. If the content is true or false. If the engagement is genuine or bought.

Traffic is traffic. Engagement is engagement.

And advertisers pay for it. Billions of dollars moving on signals that might not even be human.

## How Brands Play Along

Brands hire influencers with inflated followings. They launch campaigns built on bots. They buy ads on platforms that can't tell the difference between real eyes and fake clicks.

It's not about authenticity. It's about the appearance of reach..

Remember Fyre Festival? Sold by influencers posting one orange tile. No real product. No real infrastructure. Just hype. And it worked, until the tents and cheese sandwiches exposed the truth.

## Cultural Hits

Musicians caught inflating streams to look bigger. Influencers exposed for buying followers. Companies faking reviews on Amazon to juice sales.

And then came Cambridge Analytica. Facebook data scraped, sold, and used to manipulate voters around the world. Millions thought they were just posting family photos. In reality, their clicks were being packaged and auctioned off.

Whole industries quietly running on fake numbers.

## The Sting

We're not paying for truth. We're paying for the appearance of attention.

---

## The Human Cost

Small creators can't compete. Honest businesses get buried. Fake signals drown out real connection.

The machine rewards the loudest, not the truest. The fakest, not the realest.

And I know this firsthand.

Years ago, I had a neighbor with an account close to ten million people.

He convinced me to market with him. I paid the money. Thought it was legit.

But the results told the truth. Hollow clicks. Empty reach.

His platform was bots. Nothing more.

I've been in the game long enough to spot the difference.

And that's the trap. You think you're buying influence. What you're really buying is smoke.

How much of what you've bought online was sold to you by fake signals?

We play along. We boost posts. We chase clout. We fake it too, just to keep up.

Because the internet doesn't care who's real, it only cares who can sell the illusion.

---

**Closing One-Liner**

> *The business of fake isn't an accident. It's the economy of the internet.*

# Chapter 10 – Fake Faces, Real Influence

*Your favorite "creator" might not exist.*
*And you would never know.*

Influence used to mean people. Real people. Messy, flawed, human.

Now influence means a face. And that face doesn't even have to be real.

---

## The Rise of Digital Humans

AI influencers aren't science fiction. They're already here. Perfect skin. Perfect smiles. Perfect captions. No scandals. No mistakes. No late-night rants.

Brands love them. They never age. They never complain. They never ask for more money.

And followers? Most don't care. Some don't even notice.

We built a world where pixels can be famous.

---

## Cultural Hits

There are AI models with millions of followers. Virtual pop stars are selling out arenas. Digital avatars pushing makeup lines.

Lil Miquela was one of the first. A computer-generated it girl with brand deals from Prada to Samsung. She looked real enough. Real enough to matter. Real enough to make money.

And she won't be the last.

In China, AI livestreamers are already pulling in millions of viewers. Digital hosts run shopping streams day and night, never getting tired, never missing a cue. Thousands of people are watching, buying, and chatting without realizing the person on screen isn't a person at all.

---

**The Sting**

The most successful influencer might not even exist.

---

**The Human Cost**

It sounds harmless. A fake face selling clothes. A digital avatar posting selfies.

But it changes something deeper.

If fake people can earn real money, where does that leave the rest of us?

Creators already fight for scraps of attention. Now they're competing with code. Competing with someone who never sleeps. Someone who never burns out. Someone who can post a hundred times a day.

And the followers? They build parasocial bonds with ghosts. They laugh. They cry. They confess. They believe.

I feel it too.

My feed is flooded with pretty influencers. Maybe it's my search history. Maybe it's just the machine.

And half the time, I can't tell if they're even real.

I click on videos just to be sure. Sometimes I still don't know.

That's how far the line has blurred.

When was the last time you followed someone without knowing if they were real?

Imagine spending hours watching someone, admiring someone, maybe even loving someone, only to find out they were never real.

---

**Closing One-Liner**

*We used to say anyone can be an influencer. Now the truth is harsher: anything can.*

# Chapter 11 – Deepfakes, Scams, and Synthetic Reality

*We laugh at things that should make us cry. That's what happens when nothing feels real anymore.*

---

## The Rise of Deepfakes

At first, it was funny. Actors swapped into movies they were never in. Politicians singing pop songs.

But then the line blurred.

Now deepfakes spread faster than corrections. A fake video can circle the globe before the truth even wakes up.

And people believe it. Because video used to be proof.

---

## Scams at Scale

Voice-cloning scams are everywhere.

A mom gets a phone call. She hears her daughter crying for help. It's not her daughter. It's a cloned voice, generated from a TikTok clip.

People wire money. People panic. People fall for it.

AI scams don't need thousands of victims. They only need one.

---

## Synthetic Reality

We used to say, "pics or it didn't happen." Now pictures prove nothing. Videos prove nothing. Even phone calls prove nothing.

We're entering a reality where the fake is sharper than the real, where trust collapses not because of one lie, but because everything could be a lie.

---

## Cultural Hits

Tom Cruise deepfakes on TikTok. Millions watched a version of him that wasn't him.

Pope Francis went viral in a designer puffer jacket. A joke. A meme. But for a moment, everyone believed it.

And it goes beyond memes. An AI song mimicking Drake and The Weeknd fooled millions of listeners before it was pulled down. For a moment, the music industry shook.

Even politics isn't safe. AI-generated attack ads are already running in elections. Fake voices. Fake speeches. Fake scandals. Real consequences.

These aren't the future. These are the headlines.

---

## The Sting

The more convincing the fake, the more power it has.

---

## The Human Cost

A scam call that empties a savings account. A fake video that ruins a reputation. A deepfake that spreads faster than the truth.

And the worst part? You don't need armies of bots anymore. You just need one good prompt.

I feel it too. There was a time you could tell when something was AI. The faces were off. The hands were wrong. The voices felt stiff.

Not anymore.

Now my feeds are flooded with it. Scroll after scroll. And half the time, I can't tell what's real..

If you can't trust a photo, a video, or even a voice, what proof do you have left?

This isn't the future. It's already here.

---

**Closing One-Liner**

> *The scariest part of synthetic reality isn't that we fall for it. It's that we get used to it.*

# Chapter 12 – When the Internet is Mostly AI

*The future isn't human vs AI. It's bots vs bots, with us stuck in between.*

---

**The Flood**

The internet used to be built by people.

Now the machine is building itself.

AI writes the blogs. AI posts the comments. AI replies to the comments. AI generates the videos. AI makes the songs.

And most of it is not for you. It's for other machines. To boost signals. To farm clicks. To win the algorithm.

An endless loop of noise feeding noise.

---

**The Ghost Town Illusion**

It will look busy. Your feed will feel full.

But it will be hollow. A ghost town dressed up as a city.

Millions of posts. Millions of views. Millions of likes.

And almost none of it human.

---

## Cultural Hits

We already see it. Spam articles written by AI. Fake reviews flooding Amazon. Chatbots arguing with each other on Twitter threads.

Entire YouTube channels built by AI voices reading AI scripts over AI images. Thousands of videos. Millions of views. No people involved.

Whole news sites now run on AI, pumping out thousands of articles a day. Entire outlets where no human writer is left in the building.

TikTok and YouTube are flooded with AI-generated kids' content. Bright colors. Endless animations. Millions of views. Parents think it's harmless, but it's machines raising children on screens.

Reddit is flooded with AI-written answers. Whole comment sections generated. Whole conversations with no humans.

The future is here. Just unevenly spread.

---

## The Sting

The internet won't die in silence. It'll drown in noise.

---

## The Human Cost

If everything is content, nothing is real.

If everything is generated, what does creation even mean?

We're already losing touch. Already scrolling past ghosts. Already arguing with scripts.

And soon, most of what we see won't come from a person at all.

I felt it myself. There was a stretch when AI videos started taking over my feed. It began with the baby videos. Cute at first. Harmless.

But then it hit me. If it looks like this now, what will it be in five years?

Content creators will be a thing of the past, because all the content will be AI.

And the big companies? They'll own the machines. They'll own the future.

When you scroll, how much of what you see do you believe came from a person?

The danger isn't that we lose the internet. The danger is that we live inside it, thinking it's still alive.

### Closing One-Liner

*The end of the internet isn't empty screens. Its screens so full of fake life you forget what real looks like.*

# Chapter 13 – The Psychology of Fake

*We're not addicted to connection. We're addicted to simulation.*

---

**The Pull of Illusion**

Social media promised connection. Family. Friends. Community.

What we got was simulation. A feed that looks like life but feels like a slot machine.

Pull down to refresh. Pull down to see if someone noticed you. Pull down to feel alive for one more second.

It's not the people we're hooked on. It's the performance of people.

---

**The Dopamine Loop**

Every like. Every comment. Every view. A small hit.

It's not enough to fill us, but it's enough to keep us coming back.

We scroll, we post, we wait. The numbers trickle in. The machine tells us we matter.

And then it resets. Back to zero. Back to hunger.

---

**Cultural Hits**

The Netflix documentary *The Social Dilemma* pulled the curtain back. The very people who built the platforms admitted that what they created wasn't neutral. It was engineered to keep us hooked, engineered to feed us outrage, engineered to sell our attention back to us.

Hollywood took its own swing with *Mainstream*, a movie about manufacturing influencers out of thin air. An exaggeration, maybe, but not by much. Because online, clout is a product, and people are the packaging.

And then the headlines caught up. TikTok's own leaked documents revealed what we already felt. They decide what goes viral. They push certain content to keep us hooked, keep us scrolling, keep us numb.

None of this is an accident. Its design.

---

**The Sting**

Social media isn't designed to connect us. It's designed to condition us.

---

**The Human Cost**

We scroll not because we care what people post, but because we want to feel something.

A laugh. A distraction. A spark of life.

But it's not life. It's imitation.

The feed doesn't connect us. It rehearses connection and sells us the performance.

It gives us the illusion of being social while keeping us alone in a room.

It keeps us entertained enough to avoid silence. Numb enough to avoid ourselves.

And the scariest part is how normal that feels.

---

**Mirror Question**

Without the feed telling you who you are, would you still know?

---

**Closing One-Liner**

*We thought we were using the platforms. The truth is, the platforms have been using us.*

# Chapter 14 – The Collapse of Trust Online

*If you can't tell what's real, you stop trusting everything.*

---

## The Fracture

The internet was built on proof.

Screenshots. Receipts. Videos. Evidence.

Now none of it holds.

The more the feed fills with fake, the less we believe anything at all.

---

## When Truth Loses Its Grip

Bots flood conversations. Fake reviews flood stores. AI floods the news.

We see a headline and question it. We see a video and doubt it. We see a comment and wonder if anyone is even behind it.

Truth used to compete with lies. Now it competes with noise.

And noise always wins.

---

## Cultural Hits

Elections clouded by fake accounts and fake ads.

Movements hijacked by bots until no one remembers what was real and what was manufactured.

The Cambridge Analytica scandal showed how personal data was weaponized to manipulate voters. That was years ago. Today the tools are sharper. Faster. Cheaper.

And now political campaigns openly test AI-generated ads. Fake speeches. Fake images. Fake scandals. Real consequences.

When reality itself is flexible, trust collapses.

---

## The Sting

A society without trust cannot hold.

---

## The Human Cost

We don't just lose trust in headlines. We lose trust in each other.

Did she really write that? Did he really say that? Did they really believe that?

We start doubting everyone. Friends. Family. Leaders. Strangers.

And the cracks spread.

Because when everything feels artificial, even truth starts to sound fake.

I feel it when I watch the ape videos.

They talk. They review products. They wander through places I know.

For a moment, I believe it.

For a moment, I forget it's fake.

And then it hits me.

If it looks this real now, what happens when it gets better?

What happens when we can't tell at all?

If fake can feel that real, then trust itself is on borrowed time.

---

**Mirror Question**

If you doubt everything you see online, how long before you start doubting the people right in front of you?

---

## Closing One-Liner

*The collapse of trust online isn't just digital. It seeps into real life until nothing feels solid anymore.*

# Chapter 15 – Fame Is Not Fortune

*Social media didn't build an economy. It built a casino.*

---

## The Viral Payday

Everybody talks about going viral.
Nobody talks about the math.

A million views on TikTok? Maybe twenty bucks from the Creator Fund.
A YouTube hit? A few thousand if the ads line up.
A sponsored post? Sometimes, nothing more than free merch and a discount code.

The viral payday isn't a paycheck. It's a lottery ticket.
And the odds are brutal.

A TikTok star went viral with millions of views, only to later admit she made only a few dollars. One viral moment, and nothing to show for it. That's the economy we're being sold.

---

## The Real Math

It takes between $40,000 and $60,000 a year just to survive in most cities.
Rent. Food. Health insurance. Taxes.

Do the math.
At TikTok rates, you'd need to go viral hundreds of times just to cover the basics.

At YouTube rates, you'd need consistency that almost no one can sustain.

How many times can you hit the jackpot before you run out of luck?

---

## The Selling of the Dream

Still, the dream is sold every day.
Influencers telling kids they can quit school, quit jobs, and live off content.

But here's the secret.
Most make more money teaching the dream than living it.
Courses on how to grow your following.
Coaching on how to get brand deals.
Guides on how to go viral.

Selling the shovel is safer than digging for gold.

And some aren't even selling shovels. They're selling illusions, like the influencer exposed for renting a mansion, posing with luxury cars, and pretending it was their life. Fame becomes a costume, and the kids watching believe it's real.

---

## The Burnout Economy

Even the few who make it big are trapped. Every post is a gamble. Every algorithm shift feels like the rug getting pulled out.

Take a week off and the feed punishes you. Lose momentum and you disappear. There's no sick leave, no health insurance, no retirement plan.

And then there are the hours. Editing video after video. Chasing trends before they fade. Scheduling posts at just the right time. Filming the same clip ten times because the lighting was off.

When you add it up, the math is brutal. For what most creators actually earn, you'd be better off with a regular job. At least that comes with benefits, steady pay, and days off.

Start a small business. Sell something real. Anything beats depending on whether strangers decide to double-tap your breakfast picture.

Influencing isn't a career. It's unpaid overtime dressed up as an opportunity.

---

**The Numbers Don't Lie**

Here's the truth buried under the highlight reels:

- Over half of creators earn less than $15,000 a year.

- Only 12% of full-time creators make more than $50,000.

- Almost half earn less than $1,000 total.

- Only 4% globally ever cross six figures.

And even within those numbers, the pay is uneven. Black creators have been shown to make about a third less than white creators for the same work. Male influencers, on average, make 30% more per post than female influencers.

And when influence is tested in the real world, the numbers fall apart. One influencer with millions of followers tried to sell a new clothing line but couldn't move thirty-six t-shirts. Millions of likes, and not even three dozen sales.

This isn't an economy. It's exploitation dressed up as inspiration.

---

**The Harsh Truth**

Fame is not fortune.
Followers are not income.
Virality is not stability.

For every influencer flexing wealth, thousands are broke behind the screen.
The truth is simple: likes don't pay rent.

## Mirror Question

How many times would you need to go viral just to live?

## Closing One-Liner

> *The feed makes fame look rich. But most of it is just broke with better lighting.*

# Chapter 16 – Beyond the Fake Feed

*We don't need more content. We need more real.*

---

### The Illusion of Escape

Every now and then, people try to step away.

Delete the apps. Pause the accounts. Announce a break.

They say it's for their mental health. For focus. For family. For sanity.

And for a while, it works. The silence feels clean. The air feels lighter. The world feels bigger again.

But most come back.

Not because they missed people. But because they missed the feed. The validation. The numbers. The feeling of being seen.

The machine is too loud to ignore. Too built into work, into relationships, into culture.

Even leaving has become content. Screenshots of deleted apps. Posts about taking time off. Stories about being offline.

The exit itself becomes another entrance.

Quitting has become another form of content.

---

**Cultural Hits**

The "digital detox" movement turned into content of its own. People film themselves deleting apps, only to post about it later.

Flip phones made a comeback, marketed as an antidote to endless scrolling. But even those who switched found themselves curating the story of their switch.

Whole communities formed around quitting social media. And then those communities became their own kind of social media.

Celebrities announce they're leaving social media for good. Ye. Ed Sheeran. Dozens more. Every time, the headlines scream their exit. And every time, the return comes just as loudly.

Even our attempts at being real turn into performances.

---

**The Sting**

The fake feed isn't just on the screen. It's rewritten how we live.

## The Human Cost

We forgot how to talk without recording it.

We forgot how to live without documenting it.

We forgot how to sit in silence without reaching for something glowing in our hands.

And when we do try, it feels awkward. Unnatural. Boring.

That's how deep the fake has gone.

We don't escape the feed. We carry it with us, even when the screen is off.

## Reflective Pause

Beyond the feed, there should be life. But most of us wouldn't know what to do with it.

We're so conditioned to perform that even unplugging feels like something to share.

Silence feels like failure. Privacy feels like invisibility.

Maybe there's something real out there. But if we found it, would we even recognize it?

## Closing One-Liner

*The feed isn't just on our screens anymore. It's in our heads. And there's no logging out of that.*

# Chapter 17 – Shut It Down

*We lived before the feed. We can live after it.*

---

## Life Before the Feed

There was a time when life was not performed.
When pictures stayed in photo albums.
When conversations ended when you hung up the phone.
When silence was normal.
When boredom was part of being alive.

Then the feed arrived.

At first it felt harmless.
A way to connect.
A way to share.
A way to stay in touch.

But connection turned into performance.
Sharing turned into selling.
And staying in touch turned into never turning it off.

Now the feed isn't a tool.
It's the environment we live in.
And it's poisoning us.

---

## The Hard Reset

One option is simple.
Shut it down.

End the experiment.
Pull the plug.
Take the machine out of our hands before it rewires
us completely.

No tweaks.
No filters.
No warnings on posts.

Just silence.

It sounds impossible.
But we've done the impossible before.

We banned cigarettes in public places when the
evidence became undeniable.
We tore lead out of paint, asbestos out of walls,
poison out of products.

And even in tech, giants have fallen.
MySpace was once untouchable. Yahoo too. AOL
before that.
What feels permanent is not.

The platforms will scream collapse.
They'll say we cannot live without them.
But we did for thousands of years.
And we can again.

## The Controlled Burn

The other option is restraint.
Not erasure, but limits.

If the platforms remain, they must be broken apart.
Regulated.
Stripped of the power to decide what we see and
who we become.

We limit toxins in food.
We limit speed on roads.
We limit weapons in hands.

Why not limit the most addictive feed ever created?

Other nations already test the idea.
China locks youth gaming behind curfews and time
caps.
Europe forces tech giants into courtrooms, fining
them for privacy violations and monopoly power.
Even in the U.S., lawsuits pile up against Facebook,
TikTok, and Instagram for harming children.

The cracks are showing.
Regulation isn't a fantasy. It is already in motion.

The machine will resist.
It always does.
But unchecked, it will devour more than attention.
It will devour trust.

## The Truth

Whether it's the hard reset or the controlled burn,
one thing is clear:
We can't keep scrolling as if this is normal.

## Bold Prediction

If nothing changes, the collapse will not come from
silence.
It will come from noise.

Noise so overwhelming that truth itself will vanish.
Noise so endless that reality will fracture into a
thousand simulations.

Elections decided by bots outshouting humans.
Markets swayed by reviews no person ever wrote.
Friendships replaced by AI personas wearing
human masks.
Children raised by voices that never belonged to real
people.

When crisis comes, whether it's war, disease, or
disaster, we will not know which images to believe,
which leaders to trust, which voices to follow.

History will not be written by winners or losers.
It will be written by algorithms.

Revised. Recycled. Reinvented until nothing remains of what actually happened.

That's how societies collapse.
Not with fire.
Not with silence.
But with noise so thick it drowns out truth itself.

And here's the question every reader must face:
**When the collapse comes, will you even know which side of it you're on?**

Because when that day comes,
it won't be social media that dies.
It'll be us.

# Before You Scroll Again

This isn't the end.
The feed is waiting for you.

Before you open it, ask yourself:

- What would your life look like without the feed?

- Who are you when nobody is watching?

- How much of what you share is performance, and how much is truth?

- If the numbers disappeared tomorrow, would you still create?

- If truth disappeared tomorrow, how would you find it?

- If the feed vanished tonight, what part of you would vanish with it?

They control the feed.
You control what's real.

www.ingramcontent.com/pod-product-compliance
Lightning Source LLC
Chambersburg PA
CBHW052142270326
41930CB00012B/2983